CAR

…THING YOU NEED TO KNOW - NEW OR USED

Published By
Martin Jeffries Ltd
Dalwood House
PO Box 62
Prestwich
Manchester M25 5RG

First impression 1991

ISBN
09517183 04

Distributor
Edward & Bryning
Charles House
Royle Barn Road
Castleton
Rochdale OL11 3DT

CONTENTS

1 THE FACT-FINDING EXERCISE

So, you have decided to buy a new car. First you must decide on what car and what price range you intend to look at. *REMEMBER* - ***next to your home, this will probably be the most expensive purchase you will ever make.***

It is always best to know exactly what you are looking for: the model, the make, the price. Decide how much you want to spend and keep to it!

Before talking to the car salesman try and make a short list of possible cars that you may consider and wish to inspect. An ideal time to inspect a car is as late in the evening as possible. At this stage the sales staff will be more interested in the crossword in the evening paper than to try and close you on a deal!

This will give you time to look at the car, without feeling intimidated or under pressure by the car salesman.

If it is not possible or convenient in the evening, early Saturday or Sunday morning are perhaps the next best.

TAKE YOUR TIME, AND LOOK!

ON THE TELEPHONE

It would not be unique for you to negotiate a sale entirely by means of the telephone - without ever seeing the salesman, or the car, at least until the time of delivery. But the telephone is the most common form of initial contact, and an important and quotation gathering device.

Hopefully, by asking the right questions the answers given should give you a rough idea of the car on offer. From the details obtained you can then decide whether or not the car is a real possibility. By doing this exercise several times, a short list of vehicles can be drawn up quickly.

Remember to be confident and not aggressive to the salesman. Become assertive and not complacent, to ensure that you gather all the information you require.

A useful and convenient method of drawing up a short list of possible cars, without wasting valuable time, is by using the telephone. Before phoning draw up a list of the questions you wish to ask.

After you have established contact by telephone with the saleman it is advisable to ask for that specific contact, by name, when entering the showroom.

As you have already built up a repartee with the salesman this ease of communicating should be taken full advantage of when negotiating the deal. You are probably fully aware yourself that people prefer to do business when they feel comfortable, and are more susceptible to give that little bit extra to people they like.

So talk to the salesman, become his friend!!!

Although it would not be unusual to leave a showroom with a car on the first visit, I would strongly advise you against this. You should treat the first visit as a fact-finding mission, leaving with only a brochure, a test drive, price list and a quotation for the car you have selected.

The first rule is to forget about any part-exchanges, low rate finance, or any special additions that the dealer may have on offer. Examples to illustrate this are: A free sunroof, a radio cassette, a free 2 year warranty. Perhaps the most well-known is the £1000 offer for your old car, regardless of its condition.

USING YOUR APPRAISAL SHEET

You can record all information about the car as well as details quoted by the salesman with regard to its warranty, finance and any work he has agreed to undertake as part of the deal. So let's see what that car is really worth and how much you can buy it for.

NOTES

Make notes of questions and answers.

2 DISCOUNTS

Most main dealers work on a profit margin of 12% to 20% on the basic price of the car. The best method of achieving maximum discounts is to play one dealer against another. This can be done simply by the telephone, quoting each competitive price to the other, until you have achieved the best discount.

Before going into details about discounting, let's look at the standard pricing of a car:

We will use the hypothetical models 'TL' and 'TLX' for our example.

PRICE LIST

MODEL	*BASIC PRICE*	*CAR TAX*	*VAT*	*TOTAL*
TL	£6000	£518.27	£1140.69	**£7658.96**
TLX	£6325.08	£527.09	£1199.12	**£8051.29**
METALLIC PAINT	£152.51	£12.71	£28.91	**£194.13**

NOTE: The basic price of the car will increase, along with the car tax, on the higher specification models. Thus, it follows that model 'TLX' will be more expensive than model 'TL'.

Below shows an example of model 'TL' with a delivery charge of £200.

BASIC PRICE	**£6000.00**
CAR TAX	**£518.27**
DELIVERY	**£200.00**
VAT @ 17.5%	**£1175.69**
ROAD TAX	**£100.00**
TOTAL	**£7993.96**

If we look more closely at the deal there may be ways of misunderstanding the salesman with regard to the term "Discount". The main area of concern is: at what stage is the discount deducted from the vehicle's price?

A vital fact to remember is that the discount is always deducted from the basic price of the car, and not off the final price. An example to illustrate this is as follows:

Tabulated below are examples of two dealers both offering a 15% discount, on our hypothetical model TL.

However, as outlined, the ambiguous dealer is not actually giving the full discount being offered.

DEALER 1.

BASIC PRICE	£6000.00
DISCOUNT @ 15%	£900.00
SUB TOTAL	£5100.00
CAR TAX	£518.27
DELIVERY	£200.00
VAT @ 17.5%	£1018.19
ROAD TAX	£100.00
GRAND TOTAL	**£6936.46**

AMBIGUOUS DEALER 2.

Ambiguous dealers may offer a discount of 15% (£900) and subtract it from the total price of the car rather than the basic price.

BASIC PRICE	£6000.00
CAR TAX	£518.27
DELIVERY	£200.00
VAT @ 17.5%	£1175.69
ROAD TAX	£100.00
TOTAL	£7993.96
LESS DISCOUNT	£900.00
GRAND TOTAL	**£7093.96**

Therefore, by deducting the discount off the final price, the salesman gains an extra £135.

By using your appraisal sheet you can clearly see the itemized order, thus avoiding any 'misunderstandings' relating to discount.

Below shows an itemized order form, illustrating the model 'TL' total price. For this example we have ordered, as a factory option, metallic paint and two accessories, firstly mudflaps and secondly the glass sunroof.

A discount of 15% has been offered by the Salesman.

(N.B. - Please see previous page for model 'TL' price list)

MODEL	**£6000.00**
BASIC	
FACTORY OPTIONS	
METALLIC	£152.51
DISCOUNT	£922.87
DISCOUNTED PRICE	£5229.63
CAR TAX	
MODEL	£518.27
METALLIC	£12.71
TOTAL	£5760.62
ACCESSORIES	
GLASS SUNROOF	£140.00
MUDFLAPS	£35.00
EXTRA WARRANTY	N/A
DELIVERY CHARGES	£350.00
NUMBER PLATES	£40.00
TOTAL	£6325.61
VAT	£1106.98
ROAD TAX	£100.00
TOTAL	**£7532.59**

Another useful fact to remember is that discounting is not restricted to just the car. Discounts can also be offered on the manufacturer's extras as shown.

Now you know what you can get that new car for - but always remember to be careful of those hidden charges. An example is the price of the number plate. I have known main dealers to charge as much as £40 per set. Both you and I know that they can be obtained for as little as £4.50 from any accessory shop.

The delivery charge is also another means of making money: so keep a careful eye on the price. Although manufacturers do have a recommended price there are dealers who may increase that charge to as much as £350.

Perhaps the best time to buy is on a Sunday afternoon. This is due to the fact that every salesman in the showroom will be fighting to get every deal, so as to impress his superiors on Monday morning. This should greatly enhance your negotiating powers. If it has been a quiet weekend, he should be especially keen to please!

It may also be the case that the Sales Manager is absent, giving you a greater opportunity to negotiate a real bargain, especially with a younger and perhaps slightly inexperienced salesman.

A final word on discounts: **Don't have a £100 phone bill just to save £10 on the car!**

NOTES

Make notes of questions and answers.

3 WHERE TO LOOK FOR A USED CAR

There are three-times as many used vehicles sold compared to new. It is the business user who dominates the new market, whereas it is the millions of private buyers who make up the used sector.

For some people, buying a used car is an expensive, but relatively trouble free exercise. They are the lucky ones! For most people, buying a used car is a source of anxiety. Coupled with the hard sell techniques of the salesmen, this results in the whole process being a nightmare.

The following pages contain information, and useful hints, that should put you in a better position to approach your purchase with greater confidence and an ability to negotiate.

I trust that you will expect, rather than enquire about not paying the full asking price.
There is no need to feel reticent by behaving this way, after all the motor trade has done nothing to correct the impression that everything is negotiable, and as you now realise they account for it in an inflated retail price.

MAIN DEALER

In general, main dealers are an excellent source of quality used cars. The vehicles for sale will normally be in a condition ranging from good to 'as new' and the mileages covered will tend to be relatively low. Usually, main dealers will only sell used cars which are under three years old and under 30,000 miles.

Furthermore, because their principal business is to supply brand new vehicles, they have a significant degree of responsibility to the manufacturer who provides the dealership. Many of these major producers now operate approved used cars schemes.

Buying from this particular shop window will give you greater legal protection and a good back-up service if anything goes wrong.

Most main dealers are now squeaky clean, afraid of the manufacturers and the loss of their franchise. However, it is not the manufacturer who is party to the contract of sale (sale of the car).
It can be argued that, if a manufacturer offers a specific warranty for their product, then there is actually a collateral contract. In other words you would be entitled to take legal action against both the manufacturer and the dealer.

USED CAR DEALER

Generally, this source is not as expensive as a main dealer. Usually, they will sell cars which the main dealer will not: three years old, with more than 30,000 miles on the clock etc. There are some very reputable used car dealers with personal service which even the main dealers are sadly lacking.

In contrast to the main dealers, many established used car dealers have no service or repair facilities on their premises. This may be a disadvantage.

However, there are still rogues in the used car trade. So if you are shopping in this particular shop window, I strongly advise you to take somebody with you who has an understanding of what is on offer, and to check what back up service and warranties are available before you commit yourself.

Ask around - First impressions are important

Investigations by the Office of Fair Trading have clearly shown that "clocking" is commonplace. By removing the "speedo" head and moving the first digit in an anticlockwise direction against its ratchet, the mileage can be turned back to whatever figure is required.
If the above procedure is carried out by an amateur the tell-tale signs are obvious: The first digit, once the mileometer starts moving, slips out of alignment.

THE PRIVATE SELLER

Be careful - this in not always the cheapest or the best way to purchase a used car. Many private sales are, in fact, motor traders selling under the disguise of the private sale. Leaving you with no warranty, and themselves without any legal obligations.

Here are a few possible ways of identifying a home trader:

- **Is the name and address on the registration book the same as that of the seller?**
- **When phoning state that you are interested in the car for sale, and wait for a reply. If they answer 'which one' then this would strongly suggest that they are in fact trading.**

Private sales make up a large proportion of all cars sold. If you are able to purchase privately it usually results in a lower price than from the other sources mentioned. However, it is important to inspect the car thoroughly and perhaps ask a friendly mechanic for his advice.

Beware of the part-time motor trader. He will sell cars in poor condition, insurance 'write-offs', anything that will make him a few extra pounds.
Genuine people, wishing to sell their car privately, will normally have some facts about the car's past history. It is these facts that you should always try and establish, so as to differentiate between the private seller and the motor trader.

Remember, there are no guarantees

AUCTIONS

This option will give you perhaps the widest selection of vehicles, but you must remember that you are competing with the professional motor trader. They will know the prices of the used cars, and can afford to take a gamble. Can you?

In the fast-moving environment of the auction, you will have little time to decide, and usually no warranty, or test facilities. Although a few auction places do offer a 'right to rejection' relating to an item such as a major mechanical fault, the car must be returned to the auction within a specified time after the sale.

An auction may be regarded as a dumping ground, so look very carefully or, better still, take an experienced mechanic with you.

Perhaps the most important document you must read at the auction is the 'condition of entry and sale'. This will tell you exactly what your rights are, and at the auction they are very few!

Listen very carefully to what the auctioneer is telling you. If a fault is mentioned and the car is to be sold without warranty, that is exactly what you will get: a car with a fault and no come back!
Be careful of the impulse buy. Although your desire may be strong, you must resist. Always set your limit for the amount you want to spend before making any bids and - Remember - KEEP TO IT!

NOTES

Make notes of questions and answers.

4 CHECKING A USED CAR

The inspection of mechanical parts, and the body's major components, are very important, although you must be careful when familarising yourself with these, as it could prove both confusing and costly.

The golden rule is to take someone along with you who knows exactly what to look for. The RAC or the AA are obvious choices. If not, a friendly mechanic with experience would be a great help.

If assistance is not at hand there are simple fault finding check lists to help you during the inspection.

A knowledgeable friend will be a great help

1 START THE ENGINE

Turn the radio off and start the engine. Listen for any noises, rattles or knocks. If these are loud and do not disappear as the engine warms up there is a strong possibility that there is a serious engine fault. The engine should 'rev' smoothly without undue vibration.

2 CHECK THE INTERIOR CONTROLS

This includes anything with a switch or knob, wipers, horn, indicators, lights, as well as the brake lights. Either ask a friend or the seller to operate the lights while you check them. Make a note of the equipment that does not work.

3 CHECK THE GEARS AND CLUTCH

Whilst stationary with the engine running, push the clutch in. It should be smooth. Move the gearstick into all gear positions; there should be little resistance or noise.
When in first gear (making sure the area in front of you is clear and the hand brake is on) gently raise the clutch and try to move off. The engine should die. If the engine revs continue to rise without the car moving, this may suggest that the clutch is slipping.

4 CHECK BRAKES

Whilst stationary, press the brakes hard and keep hold for 10 seconds - they should stay hard and not go 'spongy'. Also check the seat belts, put yours on, move off and, on a quiet road, brake gently at first and then repeat more vigorously. The car should be pulled to a halt without a shudder and in a straight line.

5 SUSPENSION

When going around a long corner the car should not sway from side to side, or make any knocking noises. The twitching or veering to one side under braking, could also be caused by suspension faults.

6 AFTER THE ROAD TEST

The car will now be warmed up, so leave the engine running and look under the bonnet. Lift off the oil filler cap; there should be no bellowing steam or smoke. Check that no oil leaks have emerged and watch for any bubbles in the expansion tank; this may mean that air is escaping from the engine and points to a blown cylinder head gasket, or cracked block.

7 ENGINE IDLES BUT STALLS WHEN ACCELERATOR DEPRESSED

Check for:
Blockage in carburettor float chamber or jets
Defective carburettor accelerator pump
Manual or automatic choke inoperative

8 ENGINE RUNS ON WHEN IGNITION SWITCHED OFF

Engine overheating/coolant level low
Incorrect ignition timing
May need a decoke
Wrong grade of petrol

9 ENGINE ACCELERATES POORLY, MISFIRES OR LACKS POWER

Dirty, worn or badly adjusted contact breaker points (not applicable to vehicles fitted with breakerless ignition)
Spark plugs badly adjusted or oiled up
Incorrect ignition timing
Engine low on compression
Blockage in carburettor jets
Defective distributor

10 ENGINE OVERHEATS OR TEMPERATURE GAUGE READS EXCESSIVELY HIGH

Coolant level low
Alternator drive belt loose or broken
Hoses split or hose clips loose
Defective thermostat
Defective water pump
Defective head gasket. Deflector blown under head gasket
Faulty temperature gauge
Faulty temperature sensor(s)
Electrically operated cooling fan defect

11 ENGINE DOES NOT REACH NORMAL OPERATING TEMPERATURE

Faulty thermostat
Faulty temperature gauge or sensor

12 BRAKES JUDDER

Badly worn brake shoes or pads
Defective suspension
Badly worn brake discs or drums

13 EXCESSIVE BRAKE PEDAL TRAVEL

Rear brake self adjusting mechanism defective
Air in system
Leak in brake system
Failure in one circuit

14 BRAKES FEEL SPONGY

Air in system
Leak in brake system

15 CAR BECOMES HEAVY TO STEER

Incorrect tyre pressures
Wheel alignment requires adjustment
Suspension fault
Power assisted steering fault

16 BRAKES REQUIRE EXCESSIVE PEDAL PRESSURE

Failure in one circuit
Seized wheel cylinders
Servo inoperative
Master cylinder defective

17 DIFFICULTY IN ENGAGING GEAR

Idle speed too high
Clutch requires adjustment
Defective gear linkage
Defective gearbox or clutch
Worn out clutch

18 CLUTCH SLIPS - CAR DOES NOT ACCELERATE WHEN ENGINE REVS INCREASE

Clutch requires adjustment
Oil on clutch disc linings
Worn out clutch

19 UNEVEN TYRE WEAR

Incorrect tyre pressures
Wheel alignment requires adjustment
Wheel balancing requires adjustment
Defective shock absorbers

20 HARD OR CHOPPY RIDE

Incorrect tyre pressures
Defective shock absorbers

21 *CAR PULLS TO ONE SIDE*

Incorrect tyre pressures or uneven tyre wear
Brake adjustment incorrect
Oil, water or brake fluid on linings
Wheel cylinder(s) seized
Wheels out of alignment

22 *CAR WANDERS*

Incorrect tyre pressures or suspension fault
Uneven loading
Wheel alignment incorrect

23 *CAR LEANS EXCESSIVELY WHEN CORNERING*

Suspension fault
Defective shock absorber

24 *CAR VIBRATES WHEN DRIVING*

Wheels out of balance
Worn shock absorbers
Drive or prop shaft out of balance

25 *IGNITION WARNING LIGHT COMES ON WHEN ENGINE IS RUNNING ABOVE IDLE SPEED*

Alternator drive belt loose or broken
Alternator not charging
Defective alternator

26 *OIL PRESSURE WARNING LIGHT COMES ON WHEN ENGINE IS RUNNING*

Oil level below 'MIN' mark on dipstick
Low oil pressure or worn engine
Faulty gauge

27 *ELECTRICAL SYSTEM INOPERATIVE*

Loose or dirty battery connections
Fuse blown
Flat battery
Faulty wire connection

28 *DIRECTION INDICATORS INOPERATIVE*

Fuse blown
Indicator relay/direction indicator switch defective

DON'T BUY IN THE DARK OR THE RAIN

ALWAYS INSPECT IN DAYLIGHT!

1. Do not buy a car in either the dark or rain. You must view the vehicle in daylight in order to make a clear judgement about the paint work.
2. Always check that name, address, engine and chassis numbers are exactly the same as quoted on the vehicle registration document.
3. If the chassis plate has been re-riveted onto the body ask why? Has the car been an insurance write-off or has the car been stolen?
4. Both pedal wear and gear stick tops act as a good indication of mileage and general condition of the car.
5. The use of a magnet may prove valuable to detect rust or filler. Simply hold the magnet against the body-work: if it does not stick you have discovered an area of rust or filler!
6. Beware of thick wide "go faster" stripes: what are they hiding?
7. When buying privately always buy from the owner's home and not at a mutually agreed rendezvous or your own home. If the seller suggests this, ask why - has he a specific reason?

Listed below are just a few useful points to look for:

A ***Stand at the corner of the car and look along the side panels; repairs and dents are much easier to spot at this angle.***

B ***Check for fresh paint on the wheel arches and sills. This would suggest a recent and not very professional respray.***

C ***By lifting up the window rubbers with a pencil, paint work can be checked. Is there new paint flaking off? This may be a sign of a recent respray.***

D ***Stand well back from the car and check that the car sits evenly on the road; if not this could be an indication that there are suspension problems, or that the car has not been re-assembled correctly after an accident.***

Be suspicious of too many new parts, particularly when inspecting the engine area. New radiators, grilles, and fresh paint on the interior could suggest the car has been involved in a serious accident.

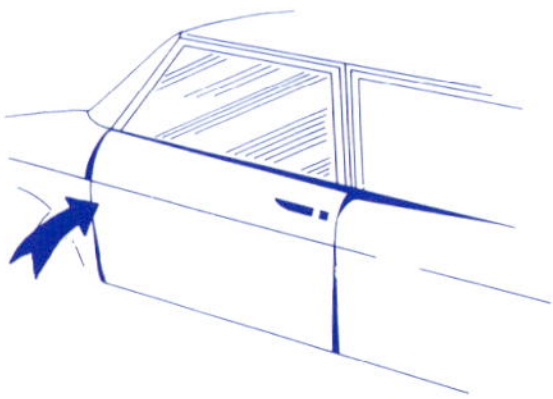

Look at the joints between panels and run your fingers between them. Are they uneven?
Ill fitting doors, bonnets, and boot-lids may suggest that the car has been re-assembled after an accident.

Amateur touch-ins should be easy to spot.

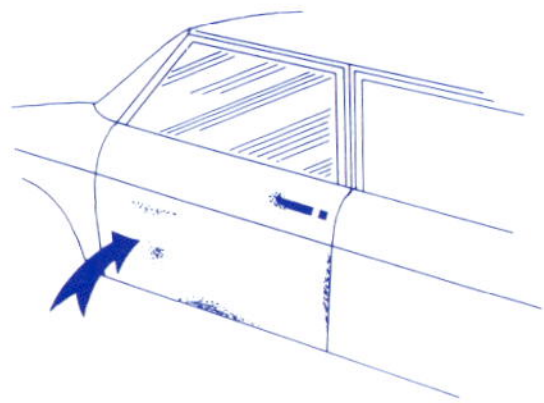

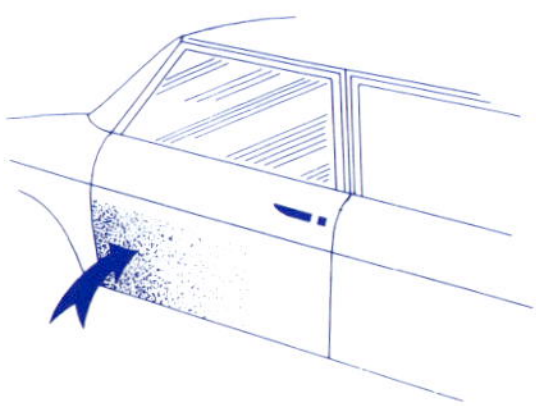

The professional way of disguising a respray is to progressively apply the new paint across the panel. However, from a distance of 5-10 feet, it should be easy to see.

From a distance, totally resprayed panels will often be a bright contrast to the exsisting paint work

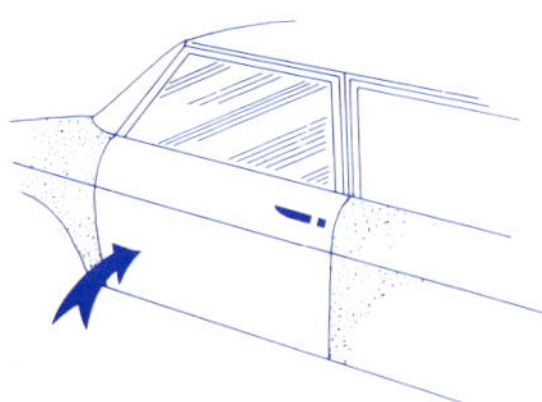

ASK FOR AN ENGINEER'S REPORT

CHECK	
ROAD SPRINGS	
SHOCK ABSORBERS	
STRUTS	
BUSHES	
CLUTCH PEDAL	
GEARBOX OPERATION	
DIFFERENTIAL	
DRIVE SHAFT	
FOOTBRAKE	
BRAKE PIPES	
FLEXIBLE HOSES	
WHEEL CYLINDERS	
HANDBRAKE	
WINDSCREEN OPERATION	
HEATER/DEMISTER	
GAUGES	
BATTERY	
STARTER MOTOR	
GENERATOR	
SIDE LIGHTS	
HEADLIGHTS	
STOP LIGHTS	
FOG LIGHTS	
INDICATORS	

Only a very few dealers display an engineer's report on their used vehicles. Until the remaining majority do, you must remain cautious about both the car and any warranty that is on offer.

If you always look at the deal from a pessimistic point of view, there should be less probability of a very expensive mistake

CONDITION

The car should be in average condition for its year and mileage. Adjustments will be made for damage, repair and mechanical faults. Bad maintenance with hard usage will be reflected in the lower trade value.

The car should look and feel right. If it is a 1986 model then that is exactly what it should look like and not something immaculate that looks as if it has just left the showroom.

If you are lucky to find a car with reasonable mileage and which has had a careful owner with the advantage of a service history, you will gain all the benefits of a good second hand car.

REGISTRATION

The year in which the car was registered is an important factor when considering valuation. The prefix registration letter changes from January to August and it is this that gives the varying value for equivalent vehicles.

An example to illustrate this is as follows:

1984 January A reg
1984 August B reg
1985 January B reg

It follows that there will be an increase in value associated with the prefix letter, i.e. B reg 1984 will have an increased value compared to that of the A reg 1984.

MILEAGE

Average mileage is estimated at 12,000 miles per year for the first year and a subsequent 8,000 miles thereafter. It follows that a high mileage vehicle will bring a lower trade value than that of an equivalent lower mileage model.

SPARES AND SERVICING

Spares and servicing can be costly when buying a prestigious or large-engine size vehicle. Although it may look like you have found yourself a bargain at the time of purchase, remember you must always take into consideration the cost of spares and/or servicing.

Another point to remember is the availability of the spares. It may be the case that a particular vehicle type has specialised parts that are not as readily available as the high volume cars such as Vauxhall and Ford.

INSURANCE

A car is placed into its respective insurance group according to the risks it presents.

The more powerful, expensive or costly to repair, the higher the insurance group and thus the higher the expense.

Group 1 is the cheapest group to insure, normally a small car with a small engine, such as the 950 Metro. Group 7 however, are far more costly and cover the larger more powerful engine cars, such as the BMW's.

FUEL CONSUMPTION

Most people have a rough idea of what to expect in terms of miles per gallon, and within the last few years good fuel consumption has become available on the high powered cars as well as the smaller sized engines.

Diesel is perhaps the most efficient source: an increase of approximately 10% in fuel consumption can be expected by the same size petrol engine.

If the car you are buying is still in production, by writing to the Department of Transport (address in appendix), they will quote the official consumption of fuel for that particular vehicle.

THE CAR

Even the colour can effect the value of the car. A cheerful red is a very sought after colour and so increases the value compared to that of beige or green, which are not in such great demand.

Small 3-door hatch backs are in greater demand than the larger engine saloons, which prove expensive to run and to buy. Thus the bigger saloons will have a lower re-sale value compared to the more popular small car.

Take sufficient time to view the car carefully and to form your own opinion on it. Do not listen to the salesman's patter.

If it feels wrong, then it most probably is wrong. If you feel comfortable and it is affordable then it's down to you to decide.

Before making a final decision on the car perhaps an independent check may be a good idea. Although you may see it as an additional cost it could well save you both worry and expense in the future, as once you have paid for the car it may prove more difficult to get your money back, should you discover a fault.

Although you should not resent the cost of the inspection, you should limit this exercise to the car you have finally chosen, as it would be foolish and costly to repeat this on every possible car.

Before you arrange for an engineer's report use your appraisal sheets, along with the information on pages 21-26, all of which can be completed by yourself. This may prove to assist the engineer in that the key areas of doubt will be already emphasized.

TEST DRIVE

TAKE YOUR TIME!
ENGINEER'S REPORT IS BEST!
SUSPICIOUS SOUNDS?
TALK TO THE SELLER!

DECISION'S TO BE MADE!
REPAYMENTS TO BE MET!
IMPULSE BUY?
VALUE FOR MONEY?
EXPENSIVE PURCHASE?

Do not always assume that the asking price is the actual price you should pay! The unscrupulous seller, whether a trader or private individual, will ask over the odds so as to leave space for negotiations, as they are well aware of the fact that they will be 'knocked down'.

DELIVERY

Always make sure that the salesman who sold you the car is present when you take delivery.

Use your appraisal sheet, even on a new car, to examine it for any damage or imperfections. It has not been unknown for a salesman to stand in front of a damaged part of the car, at the time of delivery.

Look carefully at each and every panel, and never accept a car until you are completely satisfied with its condition both inside and out.

Double Check the paper work with a calculator and never part with any money or sign any finance document until you are completely satisfied.

NOTES

Make notes of questions and answers.

5 DEALING WITH THE CAR SALESMAN

In general, the car salesman is without conscience. He will sell you a car, no matter what the price or the condition. The vehicle may have been previously used by a School of Motoring or involved in a bad accident. He may well tell you that the car has had one previous owner, but what he fails to say is that it was, in fact, the army!

Is there a good time to buy? Yes! when you feel completely confident in the deal, the salesman and the company. It is this which I feel is important: Only do business with a company and a person you like and trust. There may be exceptions to this, when it involves a real bargain. But always remember that it is you that is spending the money, so if you do not like the service do not do the business!

The salesman must sell to make a living. His only motivation is the profit margin and nothing else. The blame for this negative attitude is possibly originated by the employers: they can offer a basic salary as little as £35 per week! So you can see it is essential that the salesman sells as many cars as possible in order to achieve a realistic wage.

The salesman's wage is made from the 10-15% commission on the profit of each sale. Therefore it is a case of the survival of the fittest; no sale, no commission, no wage!

The only objective of the car salesman is to sell you a car with the maximum profit margin possible, and to give you the minimum for your part exchange (if you have one).

He will continue with his showmanship for as long as required, thinking only of the profit and his commission. His sole purpose is to get a commitment from you, the potential buyer, and he will use any tactic to get this. The most common technique of all is that of the 'closing' question; that is a question that requires a response which reduces your ability to refuse the deal. A good example of this is: "If I could, would you?"

"If I could give you £2,000 for your part exchange, would you go ahead with the purchase of the new car today?"

"If I could get your payments down to £25 per week would we have the deal?"

"If I could give you 15% discount would we have a deal now?"

The tricks will all be played, everything from giving you the impression that the unwanted cars the dealer has for sale are really like gold dust, to letting you believe the price offered is rock-bottom.

There are, however, certain things that the salesman will not be so keen to talk about and this is where the appraisal sheet becomes a valuable tool, to assist you cover these topics.

CONTROLLED SELLING

In several dealerships controlled selling is in operation. In this process, younger salesmen are under the control of one man, the 'Controller'.

He alone values and negotiates the deals. So this introduces a middle man into the deal. The very car salesman himself!

The salesman can only act as a frontman as he has no influence on the deal and therefore cannot commit himself to any specific price, or discount, or offer any estimate on the value of your part exchange.

Throughout the deal the salesman will be continually negotiating between the 'controller' and you. Therefore if a showroom does operate this system it is perhaps a useful tip, so as not to waste your valuable time, to ask to talk to the 'controller' himself.

This method of 'controlling' is usually implemented by large low category dealerships. It is the 'controller' who will calculate the cost of your instalments and inevitably they end with 99p!!

What the controller fails to tell you is that the payments are over four years at a crippling high interest rate. The object of this tactic is to remove any objection that the payments are too expensive.

NOTES

Make notes of questions and answers.

6 THE DEAL

The objective of the deal is to obtain the maximum price for your part exchange and/or to buy the car on offer for as little as possible.

The first rule is to forget about any special offers that the dealer may have and to inform the salesman that you will not be involving any part exchange in the deal, even if he saw you arrive in your car.

An example to illustrate the importance of ignoring the part exchange is as follows:

A car is on offer at £5,000. The first question to ask the salesman is "How much would the car be without any part exchange?" He may well reduce this figure by £600, giving it a true price of £4,400. Not bad for a start! Thus you will get a true picture of what, in the long run, he will be offering for your part exchange.

Once the salesman has clearly outlined the deal without your part exchange now is the time to introduce the possibility. This should clearly emphasise whether or not you are obtaining the maximum for your car.

Finding how much your car is really worth may be disappointing. This is because once discounts are introduced into the deal all part-exchanges will be bought in at a 'Trade Price'.

However, not all dealers will think alike. You may find that one may offer you more than the dealer you originally started negotiating with! This would therefore suggest that it is always wise to contact as many dealers as possible, by phone, to attain the maximum price for your vehicle.

However, an important note to be made is that you are more likely to obtain a higher value for the car if it were sold privately, thus not to involve any dealers at all!

You may wonder where the car salesman plucks that mysterious price from for your part exchange; well the answer is, the 'Book'. 'The Book', as it is called refers to either Glass's Guide or CAP. The books list prices for every model and make of car, as well as mileage considered acceptable for that particular year.

We have all heard the car salesman quote: "Well it's booked at £2,000".

For the car salesman to give you a price he must appraise and thus value the car. I recommend that you do the very same thing. I have listed points that the trade may consider important, so pay particular attention to these key areas on pages 30-31.

Due to the fact that Glass's Guide and CAP are not available to the general public it may be useful if you have a look at alternatives like Parker's Guide, which is available in the local newsagents, using them as a rough guide to the value of your vehicle.

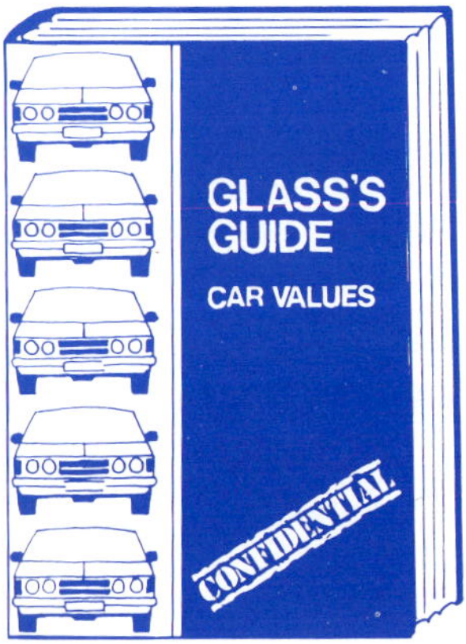

The motor trade buys and sells according to these two books, "The motor trader's Bibles".

So be prepared, the car salesman may not have as much faith in your valuation from the guides available in the local newsagents. Therefore discrepancies will arise!

Calculating how much profit a dealer is likely to make is very difficult. To establish this profit margin is extremely hard, mainly because you do not know how much he had originally paid for the vehicle.

One method of establishing a realistic idea of the vehicle's true value, is to offer the car you are thinking of buying to another dealer! By using your appraisal sheet you will have all the details to hand: mileage, year of manufacture, model etc. Therefore, by simply contacting another trader by phone and relaying the details he can quote an estimated value for the car. The price he offers will be lower than the 'selling' dealer obviously, but at least you will have attained a base mark for the car's value.

Remember, be reasonable when making the final offer. Do not be intimated and do allow the dealer to make some profit as it is his job to do so!!

This method of phoning other dealers to assess a vehicle's valuation can be extended to your own part exchange, thus obtaining a more realistic price. Therefore it may well be the case that you will receive a higher valuation from a dealer that is not involved with the purchase of your 'new' car and thus not involved with any part exchange deals.

In this way you can maximise your part exchange value, and not have the hassle of a private sale.

"HAVE I GOT A DEAL FOR YOU!"

How do you intend to pay for the car?

It may be that you have already saved the money, or that you have a bank loan arranged. Whatever the case it is highly likely that you will still be introduced to the Business Manager. It is his sole objective, like the car salesman, to sell you a package, the only difference is that this time it involves finance.

When being talked into a finance deal, be on your guard

Finance selling techniques are the same as selling the car itself. The ultimate result is your commitment to their package. The salesman that you will be introduced to may be called the Business Manager or the F & I salesman. They are both the same, F & I is simply an abbreviation for finance and insurance.

Like the car salesman they too will have an array of typical questions all of which will tend to put you in a position more likely to agree to their financial terms.

"How will you be financing the deal?"

They will always stress the importance of 'having your liquid assets for emergencies' and the fact you do not always have to use your 'first line of credit'.

Remember! always read the small print and if you are the slightest unsure take the details away with you, this will give you the time to think without the patter of the salesman.

Specific advice on finance deals is difficult as they are all tailored to the individual sale. However, research into the particular finance package you have been offered is paramount and its importance cannot be emphasised enough.

Once the package has been fully outlined by the F & I salesman compare the interest rates offered to those currently available by the high street banks. Ask for a written quotation so that all the details are to hand and are easily checked.

If zero interest is offered ensure that this is applicable to the deal offered by the car salesman and not on the full RRP (recommended retail price) of the car. Although finance packages may look cheap with a flat rate of 5% always check the 'APR' annual percentage rate.

Credit insurance is another 'add-on' to the sale, and is linked to the repayment of the finance. This facility will pay the loan if you are sick or made redundant. This comes as an additional extra so make sure it is what you want.

Consider the additional amount you will be paying for this insurance cover, compared to the amount of the original loan taken.

It may not be as competitive or as good a protection as you first thought.

Always check the 'APR' Annual Percentage Rate

By using your appraisal sheet you are able to note the APR (Annual Percentage Rate) and the flat rate of interest eg 12% or 15%.

You are also able to make note of the final monthly payments, with or without credit insurance, along with any credit you are asking for, as well as the interest that this credit will cost.

When being persuaded into a finance deal always be on your guard!

The dealer acts as a broker for the finance company, who will actually underwrite the deal. For this business the dealer will receive a handsome commission.

You may feel that the dealer is doing you a favour by arranging all the required details. He is not, finance is big business and the dealer can earn a very reasonable income from this. So make sure that you have collected as many quotations as possible before agreeing to even buying the car. Or perhaps even better, is to agree to have the car subject to you obtaining your own finance package.

Whichever the case, always read all contracts carefully and thoroughly.

7 LEASING/FINANCE

Deciding on whether to lease or to buy can only be your judgement but, perhaps, the single most important fact to remember is that leasing is essentially renting of a vehicle by the owner (the finance company) to the user, namely you, for a specified period, for a specified fee. This occurs without the user ever gaining ownership status, as the finance company never loses the title of the vehicle. On completion of the lease, the car is returned or, by arrangement with the finance company, you then may negotiate its sale to a third party. Any profits made on the sale may be recieved in the form of a refund of rentals. Additionally, there may also be the option to purchase the vehicle at a competitive price.

A dealer who offers credit terms must be licensed by the Office of Fair Trading, as a credit broker. As there are several types of credit facilities available, I have given an over-view of some of the possibilities that the dealer may offer.

PERSONAL LOAN
BALLOON/RESIDUAL PAYMENTS
HIRE PURCHASE
LEASE PURCHASE
CONDITIONAL SALE
FINANCE LEASING
CONTRACT HIRE

PERSONAL LOAN

When we say 'personal loans' one automatically thinks of a bank. However, this facility is also available from a dealer.

Perhaps the most important difference between a personal loan and other financial arrangements - eg Hire Purchase, is that you have ownership of the car from the start. You are actually borrowing money, as you may from a bank, therefore it is the money that is to be repayed. Unlike the other examples whereby the instalments are made for the car itself.

Therefore, as you have paid for the car in full the ownership automatically transfers to you from the start. Thus, it follows that the finance company cannot repossess the vehicle, should you not meet the instalments on the loan, as they did not have ownership of the car in the first place.

Another point to note is that by this method, as you are the owner from the beginning, you are free to sell the vehicle at any time. Obviously you will still have to meet the repayments of the personal loan, but is not illegal to sell the car.

BALLOON /RESIDUAL PAYMENTS

A Balloon Payment is an instalment which is defered until completion of the contract, when the sum is paid in full.

Interest is applied, not only to the nett amount that is financed, but to the balloon payment also. The interest paid can be clearly divided into two classes. Firstly, that of the amount which is financed and secondly, that of the balloon payment itself, which is usually of a higher interest rate.

Payments are made up of the capital borrowed, together with the two calculated interest charges. This total amount is paid over a specified period of time - i.e. 24, 36 or 48 months with the final payment being that of the 'Balloon'.

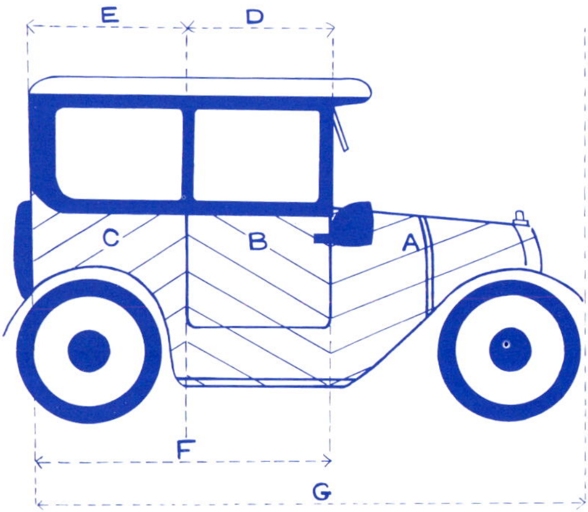

	TOTAL CASH PRICE	£10,000
A	*DEPOSIT OR ADVANCE PAYMENT*	£2,000
B	*AMOUNT FINANCED*	£5,000
C	*BALLOON RESIDUAL/PAYMENT*	£3,000
D	*INTEREST ON AMOUNT FINANCED*	£1,500
E	*INTEREST ON BALLOON PAYMENT*	£1,350
F	*TOTAL INTEREST (D + E)*	£2,850
G	*THE TOTAL PRICE*	£12,850

HIRE PURCHASE

Hire purchase is a way of purchasing a car by instalments over a period of specified time. Usually the amount of capital borrowed, together with the calculated interest changes are divided equally by the number of instalments to be paid - normally 24, 36 or 48 monthly payments.

Under the hire purchase agreement you normally, although not always, pay part of the capital cost of the car as a deposit, interest only being applied to the net amount financed.

You are essentially hiring the vehicle during the term of the contract but may exercise an option to buy, at the end of the contract time, by payment of an 'option to purchase fee'.

Ownership of the vehicle does not pass to the customer until all the instalments and the 'option to buy' fee has been paid.

LEASE PURCHASE

Lease Purchase is a modern day name for Hire Purchase or Conditional Sale: essentially there is no difference.

CONDITIONAL SALE

This differs slightly from hire purchase in that rather than hiring the car for the contract term and then taking an 'option to purchase' at the end, under the conditional terms the customer is deemed to be actually buying the car from the begining of the contract.

Ownership of the car passes to the customer on fulfilment of certain conditions, perhaps the most important is that all the instalments are paid.

With all the various types of leasing and purchasing available it follows that there will be both advantages as well as disadvantages to each method. Therefore it is essential that you consider each individual deal offered carefully.

It is always advisable to look at all options available before committing yourself. Fill in all the relevant details on the appraisal sheet so as to compare the different deals. Remember, always ask if you are unsure on any point and always seek the advice of your friendly accountant, who will explain fully the implications each option will have to your company or yourself.

BALLOON RESIDUAL/PAYMENT

Balloon payments are also available on the above.

LEASING

This is purely a financial arrangement, the same as both Hire Purchase and Conditional Sale.

The finance is paid over a period of time - i.e. 24 or 36 months. Where it does differ from both Conditional and Hire Purchase is that leasing can be seen to have two clearly defined stages, namely the primary and secondary periods.

Primary Period

Within this stage the customer pays both the finance charges and the full vehicle cost, in the form of rentals.

Secondary Period

Once the primary period has ended you may still wish to use the vehicle, this can be arranged by the payment of a low annual rental fee. This is commonly known as a "Peppercorn' rental.

Leasing can be either a fixed term or a flexible plan. In the former the payments are made over a pre-defined period of time, whereas in the latter, and possibly the most popular, the lease may be taken over a specified period initially, but throughout this time there is the option to settle the lease completely if you so desire. It is this facility that clearly sets the two forms of leasing apart, as there are no such provisions available with the fixed term lease.

So be absolutely sure that a fixed term lease is what you require as it is exactly that, a fixed lease.

Balloon Payments

Balloon Payments are also available on a lease, as illustrated earlier. This reduces the rental payments by deferring a large proportion (the Balloon Payment) until the end of the lease period.

With a balloon payment, however, you do lose the option of the secondary period - i.e. the continuation of rental.

On completion of the lease, or at the end of the primary period, or at any time within the secondary period, the customer may sell the vehicle to a third party. The profits from such a sale would be refunded to you, the customer, as a rebate of the rental.

Perhaps an important point to note is the fact that both the dealer and the finance company have responsibilities for the verbal agreements made by the salesman, as well as for the actual vehicle. This will give you that added protection and is classed as a debtor - creditor - supplier agreement.

CONTRACT HIRE

Although it is classed as a lease, the customer is in actual fact simply hiring the vehicle for a specified period together with the range of services included in the package.

This is a commercial lease arrangement, most commonly used for fleets of cars.

Thus, it can be clearly seen that this is not just another financial arrangement, but a complete motoring package.

In contract hire you have a contract for both the vehicle and a cover for its service. This will include maintenance and possibly replacement of the vehicle.

The advantage of this particular type of a rental is that it offers the large company user a complete motoring package.

Under the scheme of contract hire there are no risks, the car is simply returned to the finance company after the lease period has expired. However, it has been known for some companies to operate a maximum mileage clause whereby, if you exceed the specified mileage there will be an additional charge incurred. Another charge to watch out for is that associated with the condition of the vehicle, if it is returned in less than fair wear and tear. So be careful and read all conditions of the contract.

Therefore, it is important that you have a basic understanding of how finance leasing and the purchase of a car is defined and works within your company. I have outlined briefly the basics, but stress the importance of seeking professional advice from an accountant to assist in your individual case.

If your company makes a taxable profit, it follows that a vehicle may be purchased or leased against that profit as a legitimate business expense.

As there are so many options available, with a variety of payment schemes, I have defined the technical terms so they are hopefully more clear!

As there are so many different types, I have illustrated how lease payments can vary by tabulating some of the different kinds available. In each case the price of the car, the advance payment (or deposit), the balloon payment and interest rate are all the same - ie £10,000 for the car, a balloon payment of £3,000 and an advance payment of £2,000. However, as you will clearly see, the final instalment and the total price of the car will differ.

CONTRACT HIRE

	£
TOTAL CASH PRICE	10,000.00
SUBTRACT	
BALLOON PAYMENT/RESIDUAL	3,000.00
DEPOSIT OR ADVANCE PAYMENT	2,000.00
BALANCE TO BE FINANCED	5,000.00
INTEREST ON BALANCE - 10% x 3	1,500.00
INTEREST ON BALLOON - 15% x 3	1,350.00
TOTAL HIRE	7,850.00
MONTHLY INSTALMENTS x 36	218.05
VAT ON INSTALMENTS x 36	38.15
TOTAL MONTHLY PAYMENTS	256.20
FINAL BALLOON PAYMENT	NOT YOUR PROBLEM
TOTAL COST OF VEHICLE	11,027.00

CONTRACT HIRE WITH MAINTENANCE

MAINTENANCE	2,000.00
NEW MONTHLY PAYMENT (inc VAT) x36	321.49
TOTAL COST OF VEHICLE WITH MAINTENANCE	13,573.64

You have no car at the end of the contract!

LEASE/HIRE PURCHASE WITHOUT A BALLOON

	£
TOTAL CASH PRICE	10,000.00
SUBTRACT	
BALLOON PAYMENT/RESIDUAL	N/A
DEPOSIT OR ADVANCE PAYMENT	2,000.00
BALANCE TO BE FINANCED	8,000.00
INTEREST ON BALANCE - 10% x 3	2,400.00
INTEREST ON BALLOON	N/A
TOTAL HIRE	10,400.00
MONTHLY INSTALMENTS x 36	288.88
VAT ON INSTALMENTS	N/A
TOTAL MONTHLY PAYMENTS	288.88
FINAL BALLOON PAYMENT	N/A
TOTAL COST OF VEHICLE	12,400.00
ADVANCE PAYMENT	2,000.00
BALLOON PAYMENT	N/A
TOTAL HIRE	10,400.00
VAT	N/A
TOTAL COST WITH VAT	12,400.00

LEASE/HIRE PURCHASE WITH BALLOON

	£
TOTAL CASH PRICE	10,000.00
SUBTRACT	
BALLOON PAYMENT/RESIDUAL	3,000.00
DEPOSIT OR ADVANCE PAYMENT	2,000.00
BALANCE TO BE FINANCED	5,000.00
INTEREST ON BALANCE - 10% x 3	1,500.00
INTEREST ON BALLOON - 15% x 3	1,350.00
TOTAL HIRE	7,850.00
MONTHLY INSTALMENTS x 36	218.05
VAT ON INSTALMENTS	N/A
TOTAL MONTHLY PAYMENTS	218.05
FINAL BALLOON PAYMENT	3,000.00
TOTAL COST OF VEHICLE	12,850.00

ADVANCE PAYMENT	2,000.00
BALLOON PAYMENT	3,000.00
TOTAL HIRE	7,850.00
VAT	N/A
TOTAL COST WITH VAT	12,850.00

FINANCE LEASE WITH A BALLOON

	£
TOTAL CASH PRICE	10,000.00
SUBTRACT	
BALLOON PAYMENT/RESIDUAL	3,000.00
DEPOSIT OR ADVANCE PAYMENT	2,000.00
BALANCE TO BE FINANCED	5,000.00
INTEREST ON BALANCE - 10% x 3	1,500.00
INTEREST ON BALLOON - 15% x 3	1,350.00
TOTAL HIRE	7,850.00
MONTHLY INSTALMENTS x 36	218.05
VAT ON INSTALMENTS x 36	38.15
TOTAL MONTHLY PAYMENTS	256.20
FINAL BALLOON PAYMENT	3,000.00
TOTAL COST OF VEHICLE	12,850.00
ADVANCE PAYMENT	2,000.00
BALLOON PAYMENT	3,000.00
TOTAL HIRE	7,850.00
VAT	1,373.40
TOTAL COST WITH VAT	14,223.40

FINANCE LEASE WITHOUT A BALLOON

	£
TOTAL CASH PRICE	10,000.00
SUBTRACT	
BALLOON PAYMENT/RESIDUAL	-
DEPOSIT OR ADVANCE PAYMENT	2,000.00
BALANCE TO BE FINANCED	8,000.00

INTEREST ON BALANCE - 10% x 3	2,400.00
INTEREST ON BALLOON	N/A
TOTAL HIRE	10,400.00

MONTHLY INSTALMENTS x 36	288.88
VAT ON INSTALMENTS	50.55
TOTAL MONTHLY PAYMENTS	339.43

FINAL BALLOON PAYMENT	N/A
TOTAL COST OF VEHICLE	12,400.00

ADVANCE PAYMENT	2,000.00
BALLOON PAYMENT	N/A
TOTAL HIRE	10,400.00
VAT	1,819.80
TOTAL COST WITH VAT	14,219.48

8 CARS AND THE LAW
DEALING WITH COMPLAINTS

Dealing with complaints successfully is never easy, but having a basic knowledge of the legal principles and an awareness of when and where to seek advise is important.

You may well ask: 'What does the law say?'

In some cases there will not be a clear-cut answer to this question, whilst in others the answer will be obvious. Do not, however, always assume that the dealer is right, it has been known for them to occasionally get it wrong!

Therefore, when considering a claim, you must always have all the facts concerning your purchase to hand. This may seem an obvious point, but it is surprising how often variations on the facts occur, between the dealer and the buyer!

It is also an important point to consider how these facts would stand up in Court, should your complaint proceed to that length, so be precise. Most reliable dealers, however, will make every effort to ensure a good customer relationship is maintained and therefore endeavour to prevent going to Court.

It is the dealer's aversion to bad publicity that may be your main advantage, and thus could be used in gaining a more powerful position, when discussing your complaint.

Here are some of the questions that are often asked by people about cars. I give you some of the answers below but remember, each case may turn on its own facts and if you have any difficulties you should always consider seeking proper advice - see p.64.

Buying

1. Q. What is my position if I buy from a private individual?

A. A true private individual is only liable for:

a. Selling a car when he does not own it.
b. Making a false representation or description about the car.
c. Under the Road Traffic Acts it is an offence to sell a vehicle for use on the road when it is unroadworthy - this makes it particularly important to get good advice about your purchase.

2. Q. What are the dealer's obligations?

A. His obligations are greater than those of an individual. They are set out by the Sale of Goods Act 1979. The Act provides that any car he sells must be of 'merchantable quality'. The car must also be 'reasonably fit for its purpose' - obviously the purpose for which a car is bought is for driving on the road in reasonable comfort and safety. A car must also match the description given to it by the dealer. The dealer's obligations are additional to those a private seller has. Obviously, some wear and tear will be expected when assessing the standard for second hand cars: you have to take into account how much you have paid, how old it is, how many miles it has done and its age. You must also remember that the dealer is not liable for obvious faults or those specifically pointed out to you. The car, however, must always be safe for use on the road or the dealer would be committing an offence under the Road Traffic Acts, as we have mentioned above.

The freedom of dealers to say what they like, in order to promote the sale of cars, has frequently got them into trouble of a legal nature. A person selling goods in the course of his business, for example a motor trader, must be reasonably clear in the contents of his advertisement. The Trade Description Act states that it will be an offence for a trader to make statements that he knows to be false. This means a description in either an advertisement or brochure, for a car, which results in a misunderstanding. Any person making a false claim could be prosecuted under the Act.

3. Q. What if the dealer says: "... but it says in the Conditions here"?

A. If you are a private consumer then the Unfair Contract Terms Act 1977 says that the dealer cannot restrict his obligations under the Sale of Goods Act 1979. Any notice trying to exclude those rights may amount to an offence - this is why you see the words 'your statutory rights are not affected' in guarantees.

4. Q. What do you mean by "description"? - the dealer told me before I bought the car that it was 'a really good runner'.

A. If the dealer makes a specific description about the car then this may well be a term of your contract (example: he says 'it is a 1987 Metro 1.3'). If so, and the car is not what it is said to be, then you can return the car and claim the price back - but you must do it straight away. You may be entitled to compensation.
In addition, the dealer may be committing an offence under the Trade Description Act. This applies to all descriptions.
If the description is not so clear then it may amount only to a misrepresentation. If it encouraged you to buy the car and turns out to be false then you should be entitled to compensation (example "I've driven this one and the engine is absolutely sound" - then on the way home the engine blows up).
Words like 'a nice car' do not really mean much and therefore may have no legal effect.

5. Q. *There is a man at the door saying that the car isn't mine. What do I do?*

A. Unfortunately, this can happen. A stolen car may have false number plates and identification plates fitted to it and have the registration document belonging to another car. This is one of the reasons for buying from a reputable dealer. First, he is less likely to be taken in by the rogue seller, but more importantly he will still be there to give you your money back if you have to give the car up. It may not be so easy getting your money back from a 'fly by night' dealer or from a private individual.

There may be other reasons why the person who sold you the car was not entitled to do so. It may be that the person who sold it to you had the car on lease or hire purchase. Remember the name of the registered keeper in the registration document may not be the legal owner.

It is possible to check whether a car is on hire purchase. A reputable dealer should do this but it is also possible to do it through the motoring organizations or the Citizens' Advice Bureau. If you are the first private buyer of a car that was on hire purchase you may still be able to get ownership even though the person selling the car is strictly not entitled to do so. These provisions are contained in the Hire Purchase Act 1964.

There are some other complicated provisions in the Sale of Goods Act 1979 which may have the effect of giving you lawful ownership of the car, even though it was not sold with the permission of the true owner. This is a difficult area and it is well worth taking proper legal advice.

6. Q. *What is the position if the person selling pretends not to be a dealer but is really trading?*

A. This can be difficult to prove. If the seller is a trader then his obligations are mentioned above; in addition he may commit an offence if he advertises without disclosing his true position.

The Court of Law will look at the facts of each individual case and assess the wording used, and how the layman, namely you, would interpret them. If the advertisement states a free sunroof is available with every purchase, that is exactly what you should get.

In the case of second hand cars the relevant laws become far more involved. However, the same principles hold. If a car is sold under the description of 'one careful owner', but in fact it was used for a School of Motoring, this clearly is a false description of the vehicle. Thus a court of law should have no hesitation of finding the seller guilty of wilfully misinforming the buyer.

7. Q. I am thinking of buying a car at the auction. What is my legal position?

A. You may get a good bargain but remember, your rights may be much less. Firstly, the auctioneer is acting as agent of the seller. The auctioneer is very careful to give only a limited description of the car and very often you will find that they specifically restrict any guarantees or warranties. If you buy at an auction you will not be protected by the Unfair Contract Terms Act 1977 to the same extent. The reason is that the Act says you will not be regarded as a consumer in those circumstances. Accordingly, the auctioneer will be entitled to restrict his liability if it is fair and reasonable to do so. That may well be the case because the auctioneer does not have the time and resources to investigate the condition of every vehicle he sells. Remember: some auctioneers will allow a limited period of time in which you may say you no longer wish to buy the car. This is something that you must check before you bid.

There's one born every day

The sale of goods, by auction, is complete when the auctioneer announces so with his hammer. Until he makes that announcement you may retract your bid.
If you discover there has been a ring of dealers involved in collective bidding, ie, bidding between themselves so as to inflate the price of the car, the sale of that vehicle is not enforceable. This practice is known in the motor trade as 'Trotting'.

8. Q. Should I agree to accept the dealer's warranty?

A. Some dealers will sell you an additional guarantee which may extend for one or two years. It will not cover all parts of the car and it is important for you to look at precisely what the dealer is selling.

Over the past years, it has become common for dealers to offer customers an extended warranty on new and used cars. These warranties may often contain much legal jargon that may unfortunately disguise certain aspects of the deal being offered, which in the true light of day become nothing more than a simple mechanical breakdown insurance policy.

In some cases you will not only have the initial cost of the warranty but the additional requirements that the insurer has set down. E.g., the car must be serviced every 3,000 miles (although the manufacturer recommends every 6,000 miles). If, for whatever reason, you unfortunately miss a service the warranty becomes void.

The dealer more often than not is an agent for the company selling the warranties. He may well have bought the policy for £100 and simply sells it on to you for £200, making a very nice profit!

However, the actual contract still lies between you and the insurance company. The dealer himself will not be liable. Therefore I can only emphasize the use of your appraisal sheet to obtain all details concerning the deal. Make sure you understand exactly what the salesman is offering, and if anything is ambiguous ask him to explain.

I THINK I'VE GOT A CLAIM

1. Q. What should I do?

A. You must act quickly. If the car has a really major fault which you find very shortly after delivery, you may be able to reject the car entirely. If it is a less major defect, it would be more likely that you will be offered a repair under the warranty, but again, inform the seller straight away.

It is up to you to prove your case and you will therefore need some independent evidence. You can get a report from one of the motoring organisations or from a motor engineer. You can get limited legal advice from the Citizens' Advice Bureau and further advice from solicitors or law centres.

2. Q. The garage wants to repair it but I don't trust them any more?

A. The garage will only be liable for costs which are the reasonable consequence of what they have done wrong. If you refuse them the opportunity of putting the car right then you may find it difficult to recover the extra costs of going somewhere else. If the dealer is going to repair the car tell him that you reserve your legal rights. It is a good idea to set out the details of the complaint and what you are agreeing with the garage in a letter - make sure you keep a copy.

Normally the garage is entitled to at least one opportunity of putting the car right.

3. Q. I took the car back and it is still wrong?

A. Write to the garage again and tell them that they have not repaired the fault. If the circumstances are so bad as to show that the garage has been hopelessly incompetent then you may be justified in going elsewhere straight away. It is always better to write to the garage for a second time and tell them that unless they repair the problem this time (and give them a time limit for doing so), you intend to go elsewhere and claim those costs from the first garage.

Perhaps one of the main problems that occur when seeking to make a claim against the dealer, is that of proving the fault or damage was actually present at the time of purchase and that the dealer was aware of it.

However, do not be put-off by the quick witted dealer. You must argue that the fault was inherent - i.e. - was there at the time of purchase and that it is not unreasonable for you to have over-looked the problem at that time.

4. Q. I have written the letters but the seller didn't reply. I went elsewhere and the second garage has done the repair but the seller now refuses to pay. What should I do?

A. You should consider taking legal action. You must check that you are claiming against the right person (e.g., is your claim against an individual or a limited company). When you have found the right person you should make sure that they have the means to meet your claim. You can get some useful advice regarding small claims by obtaining a booklet from your local County Court entitled Small Claims in the County Court. The thought of going to Court may be frightening but the Court staff and the people deciding your claim will usually be quite helpful. Unless the claim seems very straight forward it may well be worth getting professional advice. If the claim is above the small claims limit (presently £500) then the position is more complicated and you should seek legal representation straight away.

5. Q. I am buying the car on finance, does this affect my position?

A. Yes, it is likely that the finance company may be jointly liable with the supplier for any breach of contract or misrepresentation. These provisions generally apply if the person providing the finance pays directly to the dealer (e.g., hire purchase or conditional sale agreements are included, so incidentally can be a sale using a credit card). You will not have rights under the Consumer Credit Act if you simply borrow the money from the bank on a personal loan or overdraft and then use that to pay the dealer.

You should not just stop paying under a Finance Agreement if something goes wrong with the car. You must get advice on that since you may find that the finance company is entitled to seek repossession of the car.

6. Q. You have mentioned that the seller may have committed an offence. Why is it important to know if someone has committed an offence?

A. If an offence has been committed then your complaint will be investigated by the police or the Trading Standards Office, which should save you time and trouble. If they decide to bring a prosecution then, when the case gets to Court, the Court has power to order compensation to be paid in your favour.

> ***If a dealer is open on a Sunday for the sale of motor spares, it is an offence for a prospective customer to be given a test drive or allowed to purchase a vehicle. However, the dealer is permitted to let the customer view the cars for sale.***

7. Q. Do I have any other remedies?

A. Yes, several. If you do not want to go to Court and the dealer or garage responsible is a member of a Trade Association that Association may have an arbitration scheme. Think carefully about adopting this course. You will lose your right to go to Court and the arbitration may well be decided only on written statements. You would lose the opportunity of explaining precisely what had gone wrong and giving the true impression to the person who is deciding the case. Remember - do not confuse this trade arbitration scheme with the County Court small claims procedure. The small claims procedure provides for a proper arbitration hearing where you will have an opportunity of having your say.

Remember every case is different. The above are intended as guidelines to give you an idea of your rights generally. Remember you may be able to get advice from the Citizens' Advice Bureau or a law centre or a solicitor.

After trying your best to settle problems direct with the trader, you can take these steps:

1. Make a formal complaint

Make a formal complaint, preferably in writing, to the manager, director, other senior executive or the proprietor of the firm. Keep a copy of your letters. If your complaint involves the warranty on a new car and the dealer cannot solve the problem, contact the manufacturer's customer relations department.

2. Seek local advice

If you are dissatisfied with the reply to your formal complaint, you can seek advice from a local authority Trading Stardards (or Consumer Protection) Department, Consumer Advice Centre or Citizens' Advice Bureau. (Find their addresses in the phone book or by asking at your Council headquarters or library.) They will advise you on any further action.

3. Contact the trade association

If your complaint is against a Code trader and he can't solve your problem, write without delay to the appropriate association. The addresses are on page 91. The trade association's staff will do all they can to settle the dispute.

4. Take your case to arbitration or court

If you are still not satisfied, you can take your case to court or, in the case of a Code trader, to arbitration or court - not to both. Before deciding, seek free advice from a local consumer adviser; or you may prefer to consult a solicitor.

9 MOTOR TRADE CODE OF PRACTICE

INTRODUCTION

This Code of Practice was originally drawn up in 1976 by the Society of Motor Manufacturers and Traders (SMMT), the Retail Motor Industry Federation (RMIF) (formerly Motor Agents Association) and the Scottish Motor Trade Association (SMTA) in consultation with the Director General of Fair Trading to govern the conduct of manufacturers and dealers, in relation to the supply of new and used cars, petrol, parts and accessories and car servicing and repair. The Code embodies principles which have been observed by the majority of the industry for many years. Improvements have now been agreed which are incorporated in the revised edition (March 1981). It will be brought up to date from time to time as the occasion demands.

The principles set out are not intended to interpret, qualify or supplant the law of the land, and are not intended to be applied on non-consumer sales.

The associations concerned regard it as a duty laid on their members that they will accept the Code in its entirety. A customer who feels dissatisfied with the treatment he has received from a member will be able to submit his grievance to the conciliation and advisory service operated by the relevant association.

It should not be overlooked that the customer also has his part to play. It is only by co-operating fully with those who make, sell and service cars that the customer can get the maximum benefit from his purchase. In particular: by maintaining his car in accordance with the manufacturer's instructions; by observing any warnings given in the handbook; and by giving as much information as possible to anyone servicing it, he can ensure he gets the best possible use out of his car and his troubles reduced to a minimum.

Throughout the Code:

The term 'Manufacturer' is taken to include concessionaire or importer.

The term 'Dealer' is taken to include retail dealer, or distributor, or supplier of goods or services, whether or not franchised by a manufacturer.

NEW CAR SALES

1.1 Dealers must bear in mind that in sale of goods to consumers they are responsible under the Sale of Goods Act 1979 for ensuring that the goods are of merchantable quality and fit for the purpose for which they are required. Statements, whether oral or in writing, which are in apparent conflict with this principle must be avoided.

1.2 Manufacturers relying on dealers to carry out a standard Pre-Delivery Inspection shall provide dealers with a standard check list for the particular model of car. Dealers must ensure that these inspections are carried out properly and a copy of the PDI check list shall be given to the customer.

1.3 The car must be delivered in condition which is to the manufacturer's standard. Each car must conform fully to all legislation affecting its construction, use and maintenance. This paragraph does not affect any legal responsibilities which may be placed on manufacturers and users to ensure this.

1.4 The benefit and limitations of any treatment, over and above that already provided by the manufacturer which is recommended by the dealer in order to inhibit the growth of rust or other corrosion, should be explained to the customer.

1.5 Order forms are intended to help both parties to the contract by spelling out the terms and conditions on which business is being done. Such terms and conditions must be fair and reasonable and set out clearly, together with a statement of the circumstances under which the order can be cancelled.

1.6 All documents must be legible.

1.7 Order forms must contain details of all charges additional to the car price so that the customer may understand clearly the total price he has to pay to put the car on the road.

1.8 Manufacturers and dealers should ensure that the manufacturer's handbook relating to the model of the car being sold is available to the customer at the time of sale of the car and for a reasonable length of time thereafter.

1.9 The terms of the manufacturers' warranties should be drawn to the attention of the customer and any relevant document published by the manufacturer must be handed over to him.

Before you buy, look carefully at all documents. Find out the total cost of putting the car on the road. Code traders must tell you this on the order form

CAR MANUFACTURER'S WARRANTIES

1 Car Manufacturers' Warranties (or Guarantees, hereafter called Warranties)

2.1 A manufacturer's warranty is a simple and straightforward way for the customer to have faults of manufacture, appearing within certain times (or before the car has done a certain mileage), put right at little or no cost to the customer without the necessity of his pursuing his legal rights against the seller.

2.2 The warranty must not adversely affect the consumer's remedies against the seller under the Sale of Goods Act 1979 and must include a statement making this clear to the customer.

2.3 The warranty must include a statement advising the customer that it is in addition to any other remedies he may have under the contract of sale.

2.4 The warranty will not extend to cover defects arising from a failure by the customer to have the car serviced in accordance with the manufacturer's recommendations.

2.5 The terms of the warranty must be easily understandable, particularly in relation to any items specifically included in or excluded from its provisions. Manufacturers and dealers should also give advice to customers as to who is responsible and what to do if they have a problem regarding parts and accessories not covered by the manufacturer's warranty.

2.6 The manufacturer should permit the transfer of the unexpired portion of any warranty to a second or subsequent owner.

2.7 The manufacturer should make clear that rectification work may be done under warranty by any of his franchised dealers. Manufacturers and dealers will take steps to ensure that warranty work is carried out rapidly and effectively.

2.8 Manufacturers will operate fair and equitable policies to permit the extension of warranty in the event of a car being off the road for an extended period for rectification of warranty faults, or of the repetition outside warranty of a fault which had previously been the subject of rectification work during the warranty period.

2.9 There is no automatic right to a loan car or contribution towards hiring charges in circumstances where a customer's own car is off the road for repair under warranty. Whether there is any such right will depend on the normal legal rules relating to damages.

Always look for the Retail Motor Industry Federation logo as shown on page 79

WARRANTY WORK

2.10 Manufacturers must give clear advice to their dealers as to the circumstances in which a loan car or contribution towards hiring charges should be provided. Such guidance should take full account of the legal position and be made available to customers on request.

2.11 Where a loan car is made available, this will merely be a reasonable alternative transport rather than an exact replacement for the car which is off the road.

Most warranties exclude liability for "wear and tear". I feel I must emphasize this point, that it is extremely difficult to prove that the fault is not due to "wear and tear" so be careful!

USED CAR SALES

3.1 Used cars sold to customers must conform to legislation affecting the construction and use of cars and should, where appropriate, be accompanied by a current Department of Transport Test (MOT) Certificate.

3.2 Dealers must bear in mind that sales of used cars are subject to the Sale of Goods Act 1979 and attention is specifically drawn to the conditions of merchantable quality and fitness for purpose contained in this Act. If, however, defects are specifically brought to the attention of the customer or the customer examines the car before the contract is made there is no condition of merchantable quality as regards those specific defects or defects which that examination ought to reveal. Dealers should therefore reveal defects on an approved checklist. See **3**.5. The format and use of the checklist are determined and approved by the trade associations or the manufacturer of the car concerned. Dealers should provide all reasonable facilities to enable prospective customers or their nominees to carry out an examination of the car prior to sale, in order that any defects which ought to be revealed at the time of sale are made known to both parties.

3.3 Dealers are reminded of the legal requirement that when selling a used car subject to a printed guarantee or warranty, that guarantee or warranty must not purport to take away or diminish any rights which the consumer would otherwise enjoy in law. The warranty document must also include a statement advising the consumer that the warranty is in addition to his statutory or common law rights.

3.4 If a printed guarantee or warranty is not used, then any specific promises which the dealer is willing to make in relation to the used car should be set out in writing.

3.5 Used cars will be subject to a pre-sale inspection in accordance with an approved checklist. The checklist must be completed before the car is offered for sale and displayed in a prominent place in the car. A copy of the checklist shall be given to the customer for his retention.

3.6 All descriptions, whether used in advertisements or in negotiations regarding the sale of used cars, should be honest and truthful. Terms which are likely to be misunderstood by the customer or which are not capable of exact definition should be avoided.

3.7 Copies of relevant written information provided by previous owners regarding the history of cars should be passed on to the customer. This may include service records, repair invoices, inspection reports, handbooks and copy of warranties, as applicable.

3.8 Reasonable steps will be taken to verify the recorded mileage of a used car and dealers will use their best endeavours to obtain a signed statement from the previous owner as to the car's mileage. Dealers should pass on any known facts about an odometer reading to a prospective customer.

3.9 Unless the dealer is satisfied that the quoted mileage of a used car is accurate, such mileage should not be quoted in advertisements, discussions or negotiations or in any documents related to the supply of the used car. Where the car's mileage cannot be verified the customer should be informed. The law requires that any disclaimer used must be as bold, precise and compelling as the car's mileage reading itself and as effectively brought to the prospective customer's attention.

Before buying a used car, ask yourself:

Is the dealer a Code trader?	***YES***	***NO***
Is an inspection checklist displayed?	***YES***	***NO***
Are all the important facts about the car - its price, condition, mileage - in writing?	***YES***	***NO***

Having an MOT Certificate is no guarantee that the car you are buying is road worthy - i.e., the vehicle is in a fit condition to be driven on the road.
All vehicles must have a registration document - i.e., the log book. The person named on this document is the "registered" keeper of the vehicle, but it does not necessarily follow that they are the legal owner too.

REPLACEMENT PARTS AND ACCESSORIES

4.1 Dealers must bear in mind that in sales of goods to consumers they are responsible, under the Sale of Goods Act 1979, that the goods are of merchantable quality and fit for the purpose for which they are required. Statements, whether oral or in writing, in apparent conflict with this principle must be avoided.

4.2 Whenever goods are offered for sale a clear indication of cash price must be available to the customer.

4.3 Terms must not be used in advertisements if they are likely to be misunderstood by the customer or if they are not capable of exact definition.

4.4 With offers of promotions or trading stamps, any restrictions which are attached to sales other than cash sales must be clearly stated.

4.5 A dealer must not display any notices or make any statements which might mislead a consumer about his legal rights in relation to the purchase of faulty goods.

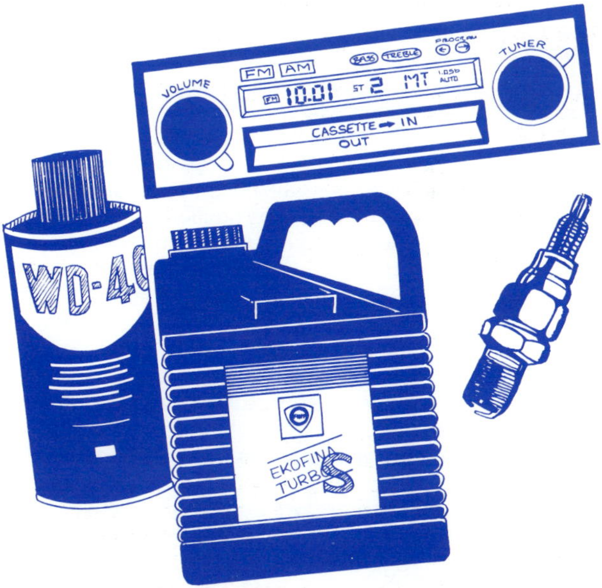

REPAIRS AND SERVICING

(Excluding work carried out under a car manufacturer's warranty)

5.1 Dealers must bear in mind that, when supplying parts or accessories in connection with repairs or servicing work for consumers, they have a similar responsibility to that which exists under a contract for the sale of goods, to ensure that the goods are of merchantable quality and fit for the purpose for which they are required and that work is performed in a proper and workmanlike manner.

5.2 Manufacturers accept a responsibility for ensuring the reasonable availability of spare parts to the distribution chain.

5.3 Spare parts should be readily available from the manufacturer from the time that a new model is offered for sale, continuing throughout its production and for a reasonable period thereafter.

5.4 Dealers will provide at least an estimate of the cost of labour and materials for all major repairs and manufacturers' recommended servicing. A firm quotation should be offered wherever possible. It must be made clear to a customer whether an estimate or quotation is being made and whether it is inclusive of VAT and where applicable the rate at which this is chargeable. Quotations should always be in writing identifying the dealer. If requested, estimates will be in writing. It should be remembered that an estimate is a considered approximation of the likely cost involved whereas a quotation constitutes a firm price for which the work will be done. If a charge is to be made for the estimate or quotation this must be made known to the customer before his instructions are accepted. Any dismantling costs which are necessary to arrive at such estimates or quotations should be notified to the customer in advance on the clear understanding whether or not dismantling costs are to be charged on an estimate or quotation which is refused. If, during the progress of any work, it appears that the estimate will be exceeded by a significant amount, then the customer should be notified and asked for permission to continue with the work.

5.5 Parts replaced during service or repair will be made available for return to the customer until the customer has taken delivery of the car unless a warranty claim is involved or unless the parts have to be submitted to the supplier because replacement parts are being supplied on an exchange basis. Dealers should notify customers in advance of work being done what their arrangements are in regard to retention and disposal of parts replaced.

REPAIRS AND SERVICING

5.6 Invoices should be clearly written or typed and give full details of the work carried out and materials used. The amount and rate of VAT should be clearly indicated. Dates and recorded mileages should always be noted where applicable.

5.7 Dealers should exercise adequate care in protecting customers' property while it is in their custody, and must not seek by disclaimers to avoid their legal liability for damage or loss. Dealers should carry adequate insurance to cover their legal liability and should strongly advise customers to remove any items of value not related to the car.

5.8 Repairs must be guaranteed against failure due to workmanship for a specific mileage or time period which should be stated on the invoice. Dealers are advised to ensure that they are adequately insured against consequential loss claims arising from any such failure.

5.9 A dealer's rules as to the method of payment he will require on completion of the work should always be notified to the customer before the work is accepted.

5.10 When it is necessary to sub-contract work, the dealer will agree to be responsible for the quality of the sub-contractors' work. Any estimate given to the customer must include the sub-contracted work and in the event of any increase in charge for the work, the principles in paragraph **5**.4 must apply.

5.11 Dealers must make it clear whether or not servicing will be carried out in accordance with the appropriate manufacturer's recommended service schedule.

5.12 While a dealer's contractual responsibility is limited to the exact terms of the customer's instructions or, for standard services, the schedule prepared by the manufacturer or other body or person, he should make it a general rule to advise the customer of any defects which may become apparent while the work is being carried out.

It is always a good idea to ask for estimates from more than one trader. But don't automatically choose the trader who submits the lowest price. Try to settle on a trader whose work you have seen or who has been recommended by friends or neighbours. Quality is just as important as price.

ADVERTISING

6.1 All advertising by manufacturers and dealers must comply with the codes and standards set by the Advertising Standards Authority and the Independent Broadcasting Authority, and with the requirements of the Trade Descriptions Act and the Energy Act (e.g. concerning fuel consumption figures). In particular, references to credit facilities must conform to the appropriate legal requirements current at the time.

6.2 Advertisements must not contain any references to guarantees or warranties which would take away or diminish any rights of a customer nor should they be worded as to be understood by the customer as doing so.

6.3 Advertisements must not contain the words 'guarantee' or 'warranty' unless the full terms of such undertakings, as well as the remedial action open to a customer, are either clearly set out in the advertisement or are available to the customer in writing at the point of sale or with the product.

6.4 Claims and descriptions in advertisements should not be misleading. In particular any comparison with other models or different manufacturers should be based on a similar set of criteria and should not be presented in such a way as to confuse or mislead the customer.

6.5 In principle, a price quoted should be a price at which the customer can buy the goods. Manufacturers and dealer should therefore quote prices for new cars, whether in advertisements or in showrooms, inclusive of the price of any extras known to be fitted to the car together with the appropriate VAT (quoting the rate applicable) and Car Tax. If the price excludes delivery charges or number plates, such exclusions must be clearly specified.

6.6 Advertisements for used cars must quote prices inclusive of VAT and they should state that the prices are inclusive.

6.7 In the description of used cars, terms likely to be misunderstood by the customer or which are not capable of exact definition should be avoided. For example, if the word 'reconditioned' is used, the nature of the reconditioning must be carefully explained.

6.8 In the description of a used car, any year must be either:

(a) the year of first use, or

(b) the year of the first registration, or

(c) the last year that the car complied with the manufacturer's specification of a model sold as new during that calendar year, whichever is the earliest.

6.9 Where an advertisement quotes the price of one model in any model range but depicts another, the actual price of that other model should also be shown.

6.10 Where a manufacturer advertise a rust-proofing process, information about the process and its limitations should be made freely available.

£10 per week
(+ VAT + interest)

£35 PER MONTH

99% APR

£3 down

Ask for written details from

0800 800 895

Due to the fact that advertisements are not always clear, it is essential that you read them carefully and thoroughly, ensuring that you have read all the small print.
It is usually contained within the small print that, e.g., Free holiday on every purchase*
****For OAP's accompanied by both sets of parents!***

HANDLING COMPLAINTS

7.1 Manufacturers and dealers must ensure as appropriate that effective and immediate action is taken with a view to achieving a just settlement of a complaint. To this end there will be, from the point of view of the customer, an easily identifiable and accessible arrangement for the reception and handling of complaints. In addition, manufacturers must give every assistance to their dealers in handling complaints under warranty, or those in which the manufacturer is otherwise involved.

7.2 When complaints are raised through a third party (e.g. the Automobile Association, the Royal Automobile Club, a Trading Standards Officer or a Citizens' Advice Bureau), willing guidance must be given to that body and every attempt should be made to re-establish direct communication with the complaining customer and to reach a satisfactory settlement with him.

7.3 In the event that a complaint in not resolved, manufacturers and dealers must make it clear to a customer that he has a right to refer the complaint to the appropriate trade association.

7.4 Manufacturers and dealers will give every assistance to the association concerned while it is investigating a complaint. The SMMT, the RMIF and the SMTA will establish liaison as necessary.

7.5 Where conciliation has failed to resolve a dispute the SMMT, the RMIF and the SMTA have agreed to co-operate in the operation of low cost arbitration agreements which will be organised by an Independent Panel of Arbitrators. Details of the arbitration arrangements are set out in the Appendix. Customers must always be advised that they have the option of taking a complaint to the Courts.

7.6 The award of the arbitrator is enforceable in law on all parties.

MONITORING

8.1 As subscribers to the Code of Practice, dealers should ensure by the clear display of appropriate symbols or other means that customers are informed of dealers' adherence to the industry's Code of Practice.

8.2 All manufacturers and dealers should maintain an analysis of justified complaints relating to any of the provisions of the Code of Practice and should take action based on this information to improve their service to the customer.

8.3 The SMMT, the RMIF and the SMTA will analyse all complaints about the Code or matters referred to the association for conciliation or arbitration. The results of such analyses will be published in the Annual Report of the relevant association.

COMPLAINTS AND ARBITRATION

1. A customer who has a complaint about the quality of the goods or services to his motor car should in the first place and at the earliest opportunity refer it to the dealer concerned.

2. The complaint, preferably in writing, should be addressed to a senior executive, a director, a partner or the proprietor. Some dealers will have an executive especially appointed to deal with complaints.

3. If the complaint relates to warranty on a new car and the dealer is unable to resolve the matter, the customer should take his complaint direct to the manufacturer concerned.

4. If attempts to reach a satisfactory solution fail, the customer has a right to refer his complaint to one of the trade associations who subscribe to the Code of Practice for the Motor Industry, if the dealer concerned is a member of that association. Any such complaint must be in writing to the addresses below.

(a) if the complaint refers to a manufacturer's warranty or lies against a dealer who is a member the address is:

The Customer Relations Department, Society of Motor Manufacturers & Traders Ltd, Forbes House, Halkin Street, London SW1X 7DS;

(b) if the complaint lies against a dealer who is a member situated in any part of the United Kingdom except Scotland the address is:

The National Conciliation Service, Retail Motor Industry Federation, 9 North Street, Rugby CV21 2AB;

(c) if the complaint lies against a dealer who is a member situated in Scotland, the address is:

Customer Complaints Service, Scottish Motor Trade Association, 3 Palmerston Place, Edinburgh EH12 5AQ.

5. All complaints referred to the appropriate trade association (SMMT, RMIF or SMTA) within a reasonable time of the cause of complaint arising will be considered.

6. If the trade association fails to resolve the complaint, its members will agree to go to arbitration except in those cases where the trade association is of the opinion that it would be unreasonable for the member to be required to do so.

7. Parties to arbitration will be asked to pay the registration fee laid down by the Chartered Institute of Arbitrators. When, later, the arbitrator makes his award, he will consider whether the registration fee should be returned to the successful party.

COMPLAINTS AND ARBITRATION

8. The parties will also be asked to sign an application for arbitration which will be sent, together with the registration fee, to:

The Independent Panel of Arbitrators, c/o Retail Motor Industry Federation, 9 North Street, Rugby CV21 2AB.

9. In order to keep costs as low as possible, the arbitration will normally rely on documents. In these cases, none of the parties to the dispute may be present nor may they be represented by any other person.

10. The trade association will submit to the Chartered Institute of Arbitrators all the documentary evidence in their possession that they consider relevant to the case. The Chartered Institute of Arbitrators will advise the parties to the dispute of the written evidence they have available on which to base their judgement and invite the parties to submit any further evidence which they consider relevant.

11. The President of the Chartered Institute of Arbitrators will appoint a single arbitrator and will make all the necessary arrangements for the arbitration to be conducted as speedily as possible.

12. In suitable cases, the Arbitrator has the right to conduct an oral arbitration and the parties may then attend to present their evidence. Legal representation may only be employed if the Arbitrator so directs.

13. The Arbitrator will have the power to direct any party to provide him and the other party(ies) any additional document or information he considers to be relevant to the matter under dispute.

14. The award of the Arbitrator will be published in writing to the parties to the dispute and to the relevant association.

15. The award of the Arbitrator is enforceable in the Courts by any party.

Be absolutely sure you clearly understand any documents you are requested to sign. If you sign a legally binding document you will be bound by it, even if you misunderstood the terms. However, if there has been a case of misrepresentation, or fraud, which leads you to believe that you were signing one type of document when in fact you were actually signing for something entirely different, then this may not be legally binding. But remember, always be careful, and read the document thoroughly.

CODES OF PRACTICE

Under the Fair Trading Act 1973, the Director General of Fair Trading has a duty to encourage trade associations to draw up Codes of Practice - rules for their members to follow which should provide a better deal and improved standards of service for customers. To show he supports a Code of Practice, a trader usually displays his trade association's symbol on his premises. Symbols may also be displayed in advertisements, leaflets and brochures.

Retail Motor Industry Federation

The advantage of Codes of Practice is that traders are expected to observe them in spirit as well as to the letter. Codes can therefore contain provisions which legislation couldn't practicably include (such as a promise to deal with complaints promptly and politely). They can be tailored to the particular circumstances and be adjusted as needs arise.

SELLING TECHNIQUES

THE ADVERT

The objective of a successful advert is to encourage the serious buyer to contact you. Having a telephone number for both day and evening would help you, of course, as would actually being in! I know it sounds obvious but, you would be suprised!

Once the potential buyer contacts you, be prepared to answer any questions about the car. Make a list, and keep it by the telephone. Here are a few examples that may be of use:

Year of manufacture
Mileage of vehicle
Length of MOT
Length of Tax
Date of last service

Or anything else that you think may be relevant.

When a person is buying a car there is always an air of indecision. So from the moment the phone rings give each potential buyer your full attention, and be prepared to negotiate around the asking price.

ASSESSING YOUR VEHICLE

Be realistic about your car. Do not ignore any faults, either mechanical or to the body work. Judge the vehicle accordingly, then decide on a realistic selling price.

The objective of the preparation stage is to make your car look clean, genuine and well maintained. A well prepared car could fetch an extra £200!

Make sure that you give the car a really good clean. Carefully touch-in any stone chips or marks on the body work. A 'touch-in' and polish, followed by a waxing should bring-up the dull paint work gleaming.

An important area which must not be over-looked is the interior; give it a really good cleaning, make use of the vacuum cleaners at many local garages. Always ensure that the windows are not forgotten, it is surprising what a difference they will make! Nasty odours in the car must also go; leave a deodoriser in the car for a few days.

If you are not in the mood for all of that work, arrange for a car valet. For around £30 it could be money well spent.

SELLING TECHNIQUES

- *Be friendly, not pushy*
- *Be polite and cheerful*
- *Emphasise the good points of the car:- new battery, service history etc*
- *Try to fix an appointment with the buyer to view the car, to confirm their commitment*
- *Be confident and always hold your ground when negotiating*
- *Have all the relevant documents readily available: MOT, Vehicle Registration*
- *Be patient, remember the buyer will always have some degree of indecision*
- *Do not give any false information about the car*
- *Remember, you will need four days' clearance on any cheques that you accept as payment for the car*
- *Better still always ask for cash*

Throughout the book we have, perhaps, taken a very pessimistic view of the car trade. This is just to ensure you are kept on your toes!!
Always take note of exactly what is being offered, whether it is a brand new Porsche or a second hand Escort.
Remember, people may not always tell you the full story, so use the appraisal sheets provided and...

GOOD LUCK!

10 HOW TO USE YOUR APPRAISAL SHEETS

Now you know the tricks of the trade you can make your position even stronger by using the appraisal sheets that follow.

The appraisal sheets, like the past chapters, are filled with questions that, accompanied with the knowledge gained from the text, should put you in a more confident position to obtain a more competitive deal.

Before you start looking for that 'new' car it is important that you have all the facts to hand regarding to your part exchange (if applicable) and what you consider a realistic value for it.

By tabulating all the details relating to your part-exchange the information becomes easily accessible for you to translate to the salesman, if and when you decide to introduce it into the deal as a trade-in.

You must also make a note of any cash you have considered putting into the deal, along with any outstanding finance settlements on your present car. To obtain a balance on any present financial settlements simply contact the bank or finance company associated with the loan.

Remember, it is important that you have all the information clearly detailed, before you intend looking for that 'new' car.

I have used a hypothetical car and figures to illustrate how to tabulate all the relevant details, so that they are to hand when the car salesman asks you for them.

Make	**FORD**	*Model*	**ESCORT**
Reg. No.	**C123 ABC**	*Date of Reg.*	**1.1.86**
Period of MOT	**10 MONTHS**	*Mileage*	**20,000**
Additional Comments	**LOW MILEAGE, ONE OWNER, RADIO/CASSETTE SERVICE HISTORY AVAILABLE**		

*Estimated Value**		**£2500.00**
Less	*Outstanding finance settlement*	**£500.00**
	Nett Value of part exchange	**£2000.00**
	Cash savings	**£300.00**
	Total advance payment	**£2300.00**

*The estimated value for your car may not be that quoted by the salesman.

Also remember you should always try purchasing the 'new' car first without involving the part-exchange, as outlined on page 39.

THE APPRAISAL

The first page of the appraisal is dual functional - i.e. - to be used with both new and used cars. This sheet is used to show brief details about the car on offer along with any external inspections carried out.

Using the diagram below you can make a note of any damage or scratches on a used vehicle, as well as using it on the collection of the new car, for the same reason.

Make	*Model*
Reg. No.	*Date of Reg.*
Mileage	*Colour*

△ ***CHIPS ETC***
□ ***RUST***
✕ ***DENT***
⁄ ***SCRATCH***

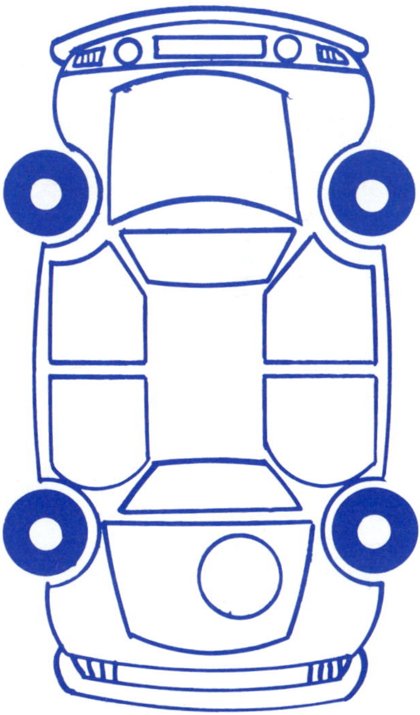

NEW CARS ONLY

This is for new cars only and as shown before, it is an itemized order form. So there can be no misunderstanding about the discount or the price you are finally paying.

MODEL	
BASIC	**£6000.00**
FACTORY OPTIONS	
METALLIC	£152.51
DISCOUNT	£922.87
DISCOUNTED PRICE	£5229.63
CAR TAX	
MODEL	£518.27
METALLIC	£12.71
TOTAL	£5760.62
ACCESSORIES	
GLASS SUNROOF	£140.00
MUDFLAPS	£35.00
EXTRA WARRANTY	N/A
DELIVERY CHARGES	£350.00
NUMBER PLATES	£40.00
TOTAL	£6325.61
VAT	£1106.98
ROAD FUND TAX	£100.00
TOTAL	**£7532.59**

TOTAL CASH PRICE **£7532.59**

USED CARS ONLY

The second page of the appraisal sheet is for used vehicles, and contains more detailed questions that may be needed to obtain any possible hidden facts about the car.

I have listed some that may be useful to ask, but emphasize the fact if you have any questions, **do ask them!**

QUESTIONS TO ASK	
Has the car been used in a school of motoring, taxi work or Hackney carriage?	
Has the car been involved in any accident or classed as an insurance write-off?	
Has the car any financial settlements outstanding?	

THOSE LITTLE HIDDEN EXTRAS!

Again on the second page I have listed some standard questions to ask the salesman, regarding to any hidden extras involved in the final purchase price.

These are to be used as a guideline to obtain a true figure for the final price.

How much is the car?	
How much is the warranty?	
How much are the accessories?	
How much is the Road Fund Tax?	
How much (if any) is the repair and/or service?	
TOTAL CASH PRICE	

FINANCIAL DETAILS

This is perhaps one of the most important parts of the whole deal and is equally one of the most difficult to understand. Because of this, it is vital that you list all details relating to both your part exchange and the vehicle you wish to purchase.

I have detailed the facts relating to your 'Trade-in', in the introduction to the appraisal section, so you should have a clear indication of the figure you have towards the final payment.

Listed below are the facts that must be completed to give you a clear indication of the final payment you will have to meet.

Total cash price of the 'new' vehicle		
Less	*Total advance payment (previously calculated)*	
	Amount of credit required	
	Charge for credit	
	TOTAL AMOUNT PAYABLE	

The 'total amount payable' is: the credit required, plus the charge for this credit.

Once you have established the total monies required, the next stage is the actual payment. Again, questions should be asked so as to attain all the factual details relating to the instalments you will have to meet.

What is the flat rate of interest? ☐

What is the A.P.R.? ☐

It may be a good idea to compare the above figures with those being offered by the High Street Banks, to ensure that you are obtaining the most competitive rate possible.

PAYMENTS

PAYMENT DETAILS

Frequently, when undertaking a loan or credit terms, a 'P.P.P.' will automatically be added to the monthly payment. This is a protective payment plan and will cover the loan payments if you are ill or are unfortunately made redundant.

Although such schemes will give you peace of mind, it is well worth asking how much this protection is costing.

Other schemes that may be added on to the payments are life insurance plans. It is always advisable to ask the salesman exactly what is included in the payments and exactly how much do these facilities cost.

It is advisable, on discussing the instalment details, that you obtain a final figure which will be exactly what is payable.

Total amount payable	
Monthly instalment for total amount payable	
Monthly P.P.P. premium	
TOTAL MONTHLY PAYMENT	
PAYABLE BY (24, 36, 48 MONTHS?)	

LEASING PAYMENT

Details of leasing on page 45

Name of Salesman
Address
Telephone Number

Notes

The back of the appraisal sheet can be used to make a note of the salesman's name and address and anything he may say about the car or deal.

11 APPENDIX

MARTIN JEFFRIES
Dalwood House
PO Box 62, Prestwich
Manchester M25 5RG

RETAIL MOTOR INDUSTRY FEDERATION
201 Great Portland Street
London W1N 6AB

THE SOCIETY OF MOTOR MANUFACTURERS & TRADERS
Forbes House
Halkin Street
London SW1 X7DS

THE SCOTTISH MOTOR TRADE ASSOCIATION
3 Palmerston Place
Edinburgh EH12 5AQ

THE AUTOMOBILE ASSOCIATION (AA)
Fanum House
The Broadway
Stanmore
Middlesex

DEPARTMENT OF TRANSPORT
2 Marsham Street
London SW17 3EB

THE ROYAL AUTOMOBILE CLUB (RAC)
RAC House
Lansdowne Road
Croydon
Surrey CR9 2JA

DRIVER AND VEHICLE LICENSING CENTRE (DVLC)
Swansea SA99 1AL

12 APPRAISAL SHEETS

I cannot emphasize enough how important it is to write down all the facts about the car, using your appraisal sheets. Once completed, the sheets will give you all the details you may need to know when comparing cars and deals offered.

MAKE NOTES!

APPRAISAL ONE

Make	Model
Reg. No.	Date of Reg.
Mileage	Colour

△ **CHIPS ETC**
□ **RUST**
╳ **DENT**
╱ **SCRATCH**

APPRAISAL ONE

MODEL	
BASIC	
FACTORY OPTIONS	
METALLIC	
DISCOUNT	
DISCOUNTED PRICE	
CAR TAX	
MODEL	
METALLIC	
TOTAL	
ACCESSORIES	
GLASS SUNROOF	
MUDFLAPS	
EXTRA WARRANTY	
DELIVERY CHARGES	
NUMBER PLATES	
TOTAL	
VAT	
ROAD FUND TAX	
TOTAL	

TOTAL CASH PRICE	

This is for new cars only and as shown before, it is an itemized order form. So there can be no misunderstanding about the discount or the price you are finally paying.

APPRAISAL ONE

QUESTIONS TO ASK	

HIDDEN EXTRAS!

How much is the car?	
How much is the warranty?	
How much are the accessories?	
How much is the Road Fund Tax?	
How much (if any) is the repair and/or service?	

TOTAL CASH PRICE	

APPRAISAL ONE

FINANCIAL DETAILS

TOTAL CASH PRICE	
LESS TOTAL ADVANCE PAYMENT	
AMOUNT OF CREDIT	
CHARGE FOR CREDIT	
TOTAL AMOUNT PAYABLE	

WHAT IS THE FLAT RATE OF INTEREST?	
WHAT IS THE A.P.R.?	

PAYMENT

TOTAL AMOUNT PAYABLE	
MONTHLY INSTALMENTS FOR TOTAL AMOUNT PAYABLE	
MONTHLY P.P.P. PAYMENT	
TOTAL MONTHLY PAYMENT	
PAYABLE BY 24, 36 , 48 MONTHS	

LEASING PAYMENT

Notes

APPRAISAL ONE

LEASING DETAILS

TOTAL CASH PRICE	

SUBTRACT

BALLOON PAYMENT	
DEPOSIT OR ADVANCE PAYMENT	

BALANCE TO BE FINANCED	

INTEREST ON BALANCE	
INTEREST ON BALLOON	

TOTAL HIRE	

MONTHLY INSTALMENTS	
VAT ON INSTALMENTS	

TOTAL MONTHLY PAYMENTS	

FINAL BALLOON PAYMENT	

TOTAL COST OF VEHICLE	

CONTRACT HIRE WITH MAINTENANCE

MAINTENANCE	
NEW MONTHLY PAYMENT (inc. VAT)	

TOTAL COST OF VEHICLE WITH MAINTENANCE	

APPRAISAL ONE

Name of Salesman
Address
Telephone Number

ADDITIONAL COMMENTS

APPRAISAL TWO

Make	Model
Reg. No.	Date of Reg.
Mileage	Colour

△ ***CHIPS ETC***
□ ***RUST***
╳ ***DENT***
⁄ ***SCRATCH***

APPRAISAL TWO

MODEL	
BASIC	
FACTORY OPTIONS	
METALLIC	
DISCOUNT	
DISCOUNTED PRICE	
CAR TAX	
MODEL	
METALLIC	
TOTAL	
ACCESSORIES	
GLASS SUNROOF	
MUDFLAPS	
EXTRA WARRANTY	
DELIVERY CHARGES	
NUMBER PLATES	
TOTAL	
VAT	
ROAD FUND TAX	
TOTAL	

TOTAL CASH PRICE [ ]

This is for new cars only and as shown before, it is an itemized order form. So there can be no misunderstanding about the discount or the price you are finally paying.

APPRAISAL TWO

QUESTIONS TO ASK	

HIDDEN EXTRAS!

How much is the car?	
How much is the warranty?	
How much are the accessories?	
How much is the Road Fund Tax?	
How much (if any) is the repair and/or service?	

TOTAL CASH PRICE	

APPRAISAL TWO

FINANCIAL DETAILS

TOTAL CASH PRICE	
LESS TOTAL ADVANCE PAYMENT	
AMOUNT OF CREDIT	
CHARGE FOR CREDIT	

TOTAL AMOUNT PAYABLE	

WHAT IS THE FLAT RATE OF INTEREST?	
WHAT IS THE A.P.R.?	

PAYMENT

TOTAL AMOUNT PAYABLE	
MONTHLY INSTALMENTS FOR TOTAL AMOUNT PAYABLE	
MONTHLY P.P.P. PAYMENT	

TOTAL MONTHLY PAYMENT	
PAYABLE BY 24, 36 , 48 MONTHS	

LEASING PAYMENT

Notes

APPRAISAL TWO

LEASING DETAILS

TOTAL CASH PRICE

SUBTRACT

BALLOON PAYMENT
DEPOSIT OR ADVANCE PAYMENT

BALANCE TO BE FINANCED

INTEREST ON BALANCE
INTEREST ON BALLOON

TOTAL HIRE

MONTHLY INSTALMENTS
VAT ON INSTALMENTS

TOTAL MONTHLY PAYMENTS

FINAL BALLOON PAYMENT

TOTAL COST OF VEHICLE

CONTRACT HIRE WITH MAINTENANCE

MAINTENANCE
NEW MONTHLY PAYMENT (inc. VAT)

TOTAL COST OF VEHICLE WITH MAINTENANCE

APPRAISAL TWO

Name of Salesman
Address
Telephone Number

ADDITIONAL COMMENTS

APPRAISAL THREE

Make	Model
Reg. No.	Date of Reg.
Mileage	Colour

△ *CHIPS ETC*
□ *RUST*
✕ *DENT*
⁄ *SCRATCH*

APPRAISAL THREE

MODEL	
BASIC	
FACTORY OPTIONS	
METALLIC	
DISCOUNT	
DISCOUNTED PRICE	
CAR TAX	
MODEL	
METALLIC	
TOTAL	
ACCESSORIES	
GLASS SUNROOF	
MUDFLAPS	
EXTRA WARRANTY	
DELIVERY CHARGES	
NUMBER PLATES	
TOTAL	
VAT	
ROAD FUND TAX	
TOTAL	

TOTAL CASH PRICE	

This is for new cars only and as shown before, it is an itemized order form. So there can be no misunderstanding about the discount or the price you are finally paying.

APPRAISAL THREE

QUESTIONS TO ASK	

HIDDEN EXTRAS!

How much is the car?	
How much is the warranty?	
How much are the accessories?	
How much is the Road Fund Tax?	
How much (if any) is the repair and/or service?	

TOTAL CASH PRICE	

APPRAISAL THREE

FINANCIAL DETAILS

TOTAL CASH PRICE	
LESS TOTAL ADVANCE PAYMENT	
AMOUNT OF CREDIT	
CHARGE FOR CREDIT	
TOTAL AMOUNT PAYABLE	

WHAT IS THE FLAT RATE OF INTEREST?	
WHAT IS THE A.P.R.?	

PAYMENT

TOTAL AMOUNT PAYABLE	
MONTHLY INSTALMENTS FOR TOTAL AMOUNT PAYABLE	
MONTHLY P.P.P. PAYMENT	
TOTAL MONTHLY PAYMENT	
PAYABLE BY 24, 36 , 48 MONTHS	

LEASING PAYMENT

Notes

APPRAISAL THREE

LEASING DETAILS

TOTAL CASH PRICE	

SUBTRACT

BALLOON PAYMENT	
DEPOSIT OR ADVANCE PAYMENT	

BALANCE TO BE FINANCED

INTEREST ON BALANCE	
INTEREST ON BALLOON	

TOTAL HIRE

MONTHLY INSTALMENTS	
VAT ON INSTALMENTS	

TOTAL MONTHLY PAYMENTS

FINAL BALLOON PAYMENT	

TOTAL COST OF VEHICLE

CONTRACT HIRE WITH MAINTENANCE

MAINTENANCE	
NEW MONTHLY PAYMENT (inc. VAT)	

TOTAL COST OF VEHICLE WITH MAINTENANCE

APPRAISAL THREE

Name of Salesman
Address
Telephone Number

ADDITIONAL COMMENTS

YOUR PART EXCHANGE DETAILS

Make		*Model*
Reg No.		*Date of reg*
Period of MOT		*Mileage*

Estimated Value		
Less	*Outstanding finance settlement*	
	Nett Value of part exchange	
	Cash savings	
	Total advance payment	

Additional Comments